Maths
Foundation

Anthology

T0173335

Published by Collins
An imprint of HarperCollins*Publishers*
The News Building, 1 London Bridge Street,
London, SE1 9GF, UK

HarperCollins*Publishers*
Macken House, 39/40 Mayor Street Upper,
Dublin 1, DO1 C9W8, Ireland

Browse the complete Collins catalogue at
www.collins.co.uk

© HarperCollins*Publishers* Limited 2021

10 9 8 7 6 5 4 3 2

ISBN 978-0-00-846891-0

British Library Cataloguing-in-Publication Data
A catalogue record for this publication is available from the British Library.

Compiled by: Peter Clarke
Publisher: Elaine Higgleton
Product manager: Letitia Luff
Commissioning editor: Rachel Houghton
Edited by: Sally Hillyer
Editorial management: Oriel Square
Cover designer: Kevin Robbins
Cover illustrations: Jouve India Pvt Ltd.
Additional text credit: Hannah Hirst-Dunton
Internal illustrations: p 2–5, 8–9 Laura Gonzales, p 6–7 Michelle Mathers, p 12–13, 18–19, 22–23 Adrita Das, p 14–15, 24–25, 30–31 Tasneem Amiruddin, p 16 Jouve India Pvt. Ltd., p 20–21 Anna Hancock, p 26–27 Q2Amedia Typesetter: Jouve India Pvt. Ltd.
Production controller: Lyndsey Rogers
Printed and Bound in the UK using 100% Renewable Electricity at Martins the Printers

Acknowledgements

With thanks to all the kindergarten staff and their schools around the world who have helped with the development of this course, by sharing insights and commenting on and testing sample materials:

Calcutta International School: Sharmila Majumdar, Mrs Pratima Nayar, Preeti Roychoudhury, Tinku Yadav, Lakshmi Khanna, Mousumi Guha, Radhika Dhanuka, Archana Tiwari, Urmita Das; Gateway College (Sri Lanka): Kousala Benedict; Hawar International School: Kareen Barakat, Shahla Mohammed, Jennah Hussain; Manthan International School: Shalini Reddy; Monterey Pre-Primary: Adina Oram; Prometheus School: Aneesha Sahni, Deepa Nanda; Pragyanam School: Monika Sachdev; Rosary Sisters High School: Samar Sabat, Sireen Freij, Hiba Mousa; Solitaire Global School: Devi Nimmagadda; United Charter Schools (UCS): Tabassum Murtaza; Vietnam Australia International School: Holly Simpson

The publishers wish to thank the following for permission to reproduce photographs.

(t = top, c = centre, b = bottom, r = right, l = left)

p 10tl Danita Delimont/Alamy, p 10tr Gabriela Staebler/Corbis, p 10c Kurit afshen/Shutterstock, p 10bl Vadim Petrakov/Shutterstock, p 10br xkunclova/Shutterstock, p 11tr Firefly Productions/Corbis, p 11tl Ernie Janes/NHPA, p 11c Terry Andrewartha/naturepl, p 11bl evenfh/Shutterstock, p 11br Erik Lam/Shutterstock, p 16cr1 Gena73/Shutterstock, p 16cl2 Ruslan Ivantsov/Shutterstock, p 16cr2 Dmitry Zimin/Shutterstock, p 16bl Gino Santa Maria/Shutterstock, p 16br Molotok289/Shutterstock, p 17tr Mega Pixel/Shutterstock, p 17c Kaspri/Shutterstock, p 17cr rvlsoft/Shutterstock, p 17bl Duplass/Shutterstock, p 17bc modustollens/Shutterstock, p 17br Alexander Baluev/Shutterstock, p 28–29 Steve Lumb

The publishers gratefully acknowledge the permission granted to reproduce the copyright material in this book. Every effort has been made to trace copyright holders and to obtain their permission for the use of copyright material. The publishers will gladly receive any information enabling them to rectify any error or omission at the first opportunity.

Extracts from Collins Big Cat readers reprinted by permission of HarperCollins *Publishers* Ltd

All © HarperCollins*Publishers*

MIX
Paper | Supporting responsible forestry
FSC™ C007454

This book is produced from independently certified FSC™ paper to ensure responsible forest management.

For more information visit:
www.harpercollins.co.uk/green

At home

I bear, I cow,
I cat – meow!

I bird, I plant.
Can you see I tiny ant?

In the garden

I can see 1 bear and 2 pears.

I can see 2 bees, and 2 birds in 1 tree.

Goldilocks and the three bears

How many bears can you see? 1, 2, 3.

How many beds? 1, 2, 3.

The stream

Show me 1 frog. Show me 4 snails.

Show me 2 flowers. Show me 5 ducks.

Show me 3 biscuits.

1, 2, 3, 4, 5. That's 5.
Show me, show me,
What else can you see?

How many animals?

1 crocodile

2 elephants

5 frogs

7 horses

8 dogs

3 lions

4 birds

6 rabbits

9 meerkats

10 cats

The fruit stall

How many kinds of fruit are there?
6 apples, 5 melons, 4 juicy pears.

3 pineapples, 2 purple plums.
I shopkeeper, I girl, I mum!

Carpet patterns

What patterns do the carpets show?
In what order do they go?

How many colours can you name?
Are the patterns all the same?

2D shapes

Circles, triangles, squares and more.
Find shapes in food, toys, signs
and doors.

Can you match shapes that look the same?

As you match, say each shape's name.

Some things are cubes. Some things are spheres.

Can you find some more shapes here?

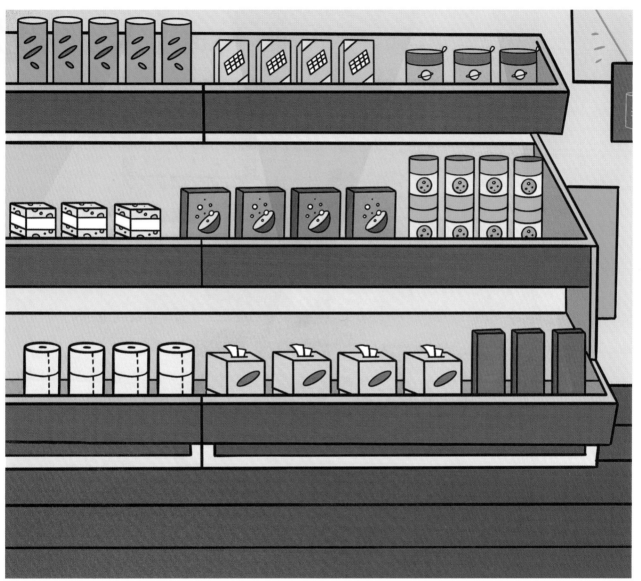

Do you know all the other shapes' names? What things are different? What are the same?

Aliya's room

In Aliya's room there are lots of cats.
They're on a table and the mat.

On the screen, in the bed,
On the wall, above her head!

At the park

Some of these things are high and tall.
Some of these things are low and small.

Which are narrow and which are wide?
Look up: which is the highest kite?

The farmyard

Look for the insects, small and light.
They float on the air as they
take flight.

Which animal is heavier, the duck or the horse?
The horse is bigger and heavier, of course!

Lee's day

time to get up

time to get dressed

story time

home time

time for school

playtime

bath time

bedtime

Luke's exercise diary

Monday

Tuesday

Friday

Saturday

Thursday

Wednesday

Sunday

Selvan's bakery

Look at the things that Selvan bakes!
Can you describe the things he makes?

What kind of thing is on each tray?
Why did Selvan make groups this way?

Numbers 1 to 10

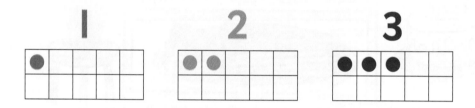

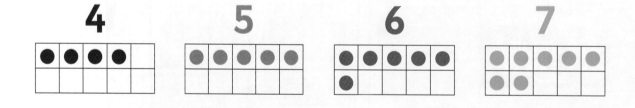

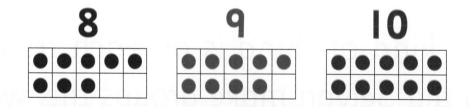